Engineering on the Edge:

The Future of Nanotechnology and Robotics

by

Jon D. Sutherland

authorHOUSE™
1663 Liberty Drive, Suite 200
Bloomington, Indiana 47403
(800) 839-8640
www.AuthorHouse.com

This book is a work of non-fiction. Names of people and places have been changed to protect their privacy.

First published by AuthorHouse 10/28/04

ISBN: 1-4184-9000-8 (sc)
ISBN: 1-4184-8999-9 (dj)

Printed in the United States of America
Bloomington, Indiana

This book is printed on acid-free paper.

TABLE OF CONTENTS

PREFACE

Should industries automate with the newest tools of technology - nanotechnology and robotics? Whether the future of industry will put into practice one of the most original forms of automation is yet to be seen but, assuredly, the engineers and scientists of the new millennium are ushering in an age of autonomous and service machines. The chosen topics of this book are nanotechnology and robotics.

Nanotechnology is molecular manufacturing, more correctly phrased as *Molecular Nanotechnology (MNT)*. This generally refers to building objects one atom or molecule at a time. In size, nanometers, or billionths of a meter, equates to a thousand, million times smaller than a meter. A nanometer is one-billionth of a meter.

Robotics is characterized here in three categories: industrial and, in particular, service and autonomous, and is presented as the method of interaction between humans and robots. Today, the thresholds set by science fiction writers for robotic implementation, once only imagined, are now becoming realities.

This book will give insight into both the future of nanotechnology and robotics, and the combining of nanotechnology and robotics. Both of these subjects in present and futuristic terms are singularly impressive, and when combined can become the frontier of engineering possibilities. With these topics, imminent advancements in automation will become realities.

The utilization of nanotechnology and robotics in future manufacturing will be a driving force if the trend for factory automation is to continue. America will have to provide the funding and management to remain competitive in the international markets. In dealing with highly competent world business competitors, survival depends upon equally competent responses. Historical analysis indicates that in the United States most investments are made in people, manufacturing processes and tools. Many economists explain the relative decline of U.S. manufacturing simply as economic evolution, with the U.S. moving toward a service economy. Without a longer-term manufacturing investment focus, America could fall behind foreign competitors in product production. International position is partially driven by the application of these new technologies, continued research and development, acceptance of U.S. and industrial standards, and fair international trade agreements.

These findings, although applied to two separate fields (nanotechnology and robotics), are and will become even more interdependent upon each other in the future. With the capabilities of nanotechnology and robotics in the manufacture of devices, robots and smaller machines, things that before now were impossible to create, have given way to possibilities and in some cases are becoming realities.

A Bookmark to the Future

Due to the incredible speed at which nanotechnology and robotics are developing, the references used and information gathered is either from the principal source or from recognized leaders in the field. Along with this, pioneers and persons who were instrumental in the melioration of nanotechnology and robotics were selected so as to preserve the integrity of the findings.

Only primary information will be emphasized to give the original intentions and scope into these engineering fields. While sometimes historical events are profiled, special attention is given to accuracy and these events present a necessary background into the present-day use of engineering with nanotechnology and robotics.

In the area of nanotechnology, in addition to books, releases and conferences from the pioneers of the subject, the World Wide Web is currently an acceptable resource so long as the source has been validated as a true authority providing facts that can and have been checked, and the authors have been recognized for their reputable work.

Due to the new and innovative nature of this subject, the books on nanotechnology are presently fewer than in robotics.

In the area of robotics, a combination of respected authors and the resources of the World Wide Web were used. The number of authors on this subject is greater than nanotechnology due to the length of time in which robotics has existed.

At the current pace of nanotechnology and robotics, this book will quickly become dated and will serve only as a bookmark to the future of these new engineering endeavors.

Discoveries Along the Way

Small, cheap, efficient and precise are the call words of today's businesses. The disproportionate rise in the service sector is bringing on a demand for a good value-to-money ratio with the goals being to increase the utilization of existing services and to bring about new fields of operations for individuals and services. Because of these findings, the fields of nanotechnology and robotics are particularly suited to fulfill these requirements.

Nanotechnology is a field of nanoscience and nanoengineering where there is the ability to manipulate and move matter. This is leading to an unprecedented understanding of the fundamental building blocks of all physical things. These developments are likely to change the way almost everything from vaccines to computers to objects not yet imagined are designed and made.

Some areas of optimization of product properties are: Energy Engineering: batteries, gas cells, gas-garrets, solar cells; Environmental Technology: cleaning, material cycles, disposal; Information Technology: efficient processors, high-density materials; Medical: health and aging developments.

Within the medical field, nanotechnology will revolutionize the way patients are given medical aid and treated for illness.

Scientists at Northwestern University have become the first to design molecules that could lead to a breakthrough in bone repair. By emulating nature in devising biomaterials that are not only biocompatible, but are also biodegradable and are nanostructured, the possibilities could lead to bone repair and counteract the effects of bone cancer, osteoporosis, and other bone related diseases. "Recreating natural bone structure at the nanoscale level - the first level of bone structure hierarchy - is what we set out to do with our experiments, and we succeeded," said Northwestern postdoctoral fellow (1999-2002) and Rice University Professor, Jeffery D. Hartgerink, the leading author of a paper reporting these results. (Jeffery D. Hartgerink joined Professor Samuel I. Stupp's laboratory at Northwestern University, Evanston, IL.)

Nanotechnology in the military represents technologies that can alter the course of battles and precipitate the rise or fall of nations. Nanotechnology has clear military and terrorist uses. The possibilities are light-weight equipment, Unmanned Aerial Vehicles (UAV) - these could provide reconnaissance, using radio signals to transmit real-time information from its camera to a display. Soldiers with micro devices integrated into their uniforms to provide Identification Friend or Foe (IFF). The Massachusetts Institute of Technology (MIT) received a grant from the army to open a Institute for Soldier Nanotechnologies (ISN). The center will help create uniforms and materials that could help protect soldiers against bullets and chemical agents or monitor their life support processes.

Robotics will give the field of nanotechnology a medium through which they both can benefit. As in the military's use of Unmanned Aerial Vehicles (UAV), Lockheed Martin created a UAV that is only 6 inches long and weighs 3 ounces.

Raytheon is currently expected to produce a military receiver that weighs 4 ounces. It will work ten times longer than current models and require less maintenance. The receiver is part of a larger effort to shrink a four-channel radio, now weighing 10 pounds, to the size of a credit card.

The Land Warrior helmet is part of a new military fighting system integrating small arms with high-tech equipment. Monocle visual display in the Land Warrior helmet shows maps, data, position, manuals, and orders from headquarters. Microtechnology makes possible a high-resolution, low power display screen (0.5 to 5 inches) meeting mobility requirements and fitting into the larger computer network.

Both of these fields will give companies an excellent chance to participate in innovative concepts. Development through these two fields of engineering is accelerating, according to experts, due to the foreseeable forecast in the service sector. Accordingly, experts in the robotics field forecast that service robots will overtake the number of industrial robots.

Secondary effects will also reduce cost. Microscale systems require less energy to operate moving parts. Systems that run on lower power produce less heat, leading to fewer maintenance problems and a longer service life. Moreover, smaller systems that weigh less require less energy to propel. Other advantages stem from the physical properties of very small devices. Many use electrostatic energy for power, drastically reducing energy requirements.

CHAPTER 1
Introduction

Will the technologies of the new millennium bring benefits or concerns? A new age of engineering genesis has begun and it is evident that new technologies and methods are being implemented throughout the world. Two of these technologies are nanotechnology and robotics. These two technologies were chosen because of the rapidity at which the technology is being pursued and how profoundly different their uses will affect the world market compared to the methods and technology of the previous century.

Nanotechnology

In the case of nanotechnology, it matches the scheme of "smaller is better." Nanotechnology is a scale and scope of which is right out of science fiction. It is the manipulation of individual atoms. The scale of a nanometer is one billionth of a meter, approximately 3 to 4 atoms wide.

According to Dr. Ralph Merkle, former Principle Fellow at Zyvex, and present Professor of Computing and Director of Georgia Tech Information Security Center (GTISC) at The Georgia Institute of Technology in Atlanta, Georgia, "Nanotechnology will replace our entire manufacturing base with a new, radically more precise, radically less expensive, and radically more flexible way of making products."

Nobel Laureate Dr. Richard Smalley, testifying before congress during a hearing on nanotechnology on June 22, 1999, said: "The impact of nanotechnology on health, wealth, and lives of people will be at least equivalent of the combined influences of microelectronics, medical imaging, computer-aided-engineering, and man-made polymers developed in this century.

Some examples of nanotechnology include manufacturing technology capable of making trillions of tiny machines, each the size of a bacteria. Each machine contains an onboard controller programmed to control a set of molecular scale tools and manipulators. An individual machine may be designed to manufacture superior materials, convert solar energy to electricity, or even, ultimately, enter the body to fight disease and aging at the cellular and molecular level. Materials hundreds of times better than today's best materials, vastly more powerful computers, precise machinery that doesn't wear out, and a revolution in clean manufacturing are but a few of the predicted benefits of applying these new machines.

Robotics

Robots are becoming more common in the workplace, and their initial task is to do much of the work in factories that is dangerous and difficult for humans. They are especially suited to do onerous, repetitive, and otherwise unpleasant tasks that are required of present-day production employees. Many of the robots in use today perform jobs that require great strength, pose danger to humans, or where humans are prone to make simple but costly errors. The decision to employ a robot is sometimes not because they are faster or cheaper than humans, but because they work in a place where humans cannot.

Presently, robots are typically introduced into the workplace in three application areas: Industrial, Special-purpose, and Autonomous.

The Industrial robot will see service as pick-and-place units, loading/unloading, assembly operations, spot welding, spray painting and inspection. Its main features are: repetitive operation, preprogrammed, rigid, structured environment.

The Special-purpose robot can be applied to nuclear material handling, underwater handling and inspection, space servicing (repair, assembly and exploration), civil engineering, and personal service. Its main features are: no repetitive tasks, mobile, flexible behavior, unstructured environment.

The Autonomous robot can find use in an industrial setting, exploration of unknown terrain, transportation and handling of tools and work pieces, inspection, automatic traffic control, military and security areas. Its main features are: no repetitive tasks, mobile, flexible behavior, learning capabilities, complex sensor systems, unstructured environment.

So as these new methods and technologies change the way business is conducted and modifies the course of lives, what will they do and how will they do it is the greater question.

Purpose of This Book

The intent here is to present specific and specialized methods and technologies which are in their infancy today and, in all likelihood, will change the way of the world tomorrow. The subjects covered are current fields to modern engineering today and present a viewpoint on advances in technology as related to the role of engineering. The chosen accent of this book is to bring to the forefront the achievements which have become outstanding as having a definite influence on society and some of the myriad of futuristic endeavors.

The Essence of Engineering

Science, technology and craftsmanship grow together. With successful stimulation of all three, further advances are achievable. From the beginning of time to the present, man has worked to make his world a better place to live in. The essence of engineering is the utilization of the resources and laws of nature to benefit humanity. Engineering is a major segment of our lives - it deals with conception, design, development, refinement, and application of machines and mechanical apparatus of all kinds. The subjects chosen, nanotechnology and robotics, will test engineers and the limits of present-day materials and methods of manufacture.

What in the World is Nanotechnology?

Nanotechnology is molecular manufacturing, more correctly phrased as *Molecular Nanotechnology (MNT)*. Simply worded, nanotechnology means building things one atom or molecule at a time with programmed nano-scale robot arms. Nanotechnology is only in the research and development stage. The long-reaching implications of nanotechnology are far beyond any present accomplishment. Some of these anticipated future accomplishments may include:

- Computers which are millions of times faster.
- Inventions which are impossible today - molecular scale versions resembling their macroscopic counterparts.
- Safe and affordable space travel.
- Medical nanotechnology - a virtual end to illness.
- Molecular food syntheses - possible end of famine and starvation.
- Reintroduction of many extinct plants and animals.
- Self-assembling consumer goods.

Nanotechnology is not only concerned with building smaller machines. It has been found that by changing certain material grain structure, this would also change the normal characteristics of those materials. One of these material changes is conductivity which can be either increased or decreased.

The applications for nanotechnology are applied to the fields of: biomedicine, electronics, engineering, medicine, optics and physics. Applications include automotive accelerators and pressure sensors, biomedical instruments, fiber optic switching systems, and inkjet nozzles.

The goal of nanotechnology will be to build machines the size of molecules. Nanomanufacturing today is very basic and at the molecular level. It has been compared to moving a child's building blocks with boxing gloves on the hands of the builder: "It's easy to move them around, but hard to build something with the blocks." (Cone, Robert J. Second Edition. Revised and Updated by Patricia Barnes-Svarney. How The New Technology Works [a guide to high-tech concepts]. Oryx Press. 1998)

The man who began to steer this field of study and is also given credit for coining the term "nanotechnology" is Dr. K. Eric Drexler. He obtained an SB and SM from the Massachusetts Institute of Technology (MIT), and was awarded a Ph.D. from MIT in Molecular Nanotechnology (the first degree of its kind), supervised by Marvin Minsky. His book

Nanosystems was awarded the AAP 1992 Most Outstanding Computer Science Book. His work on nanotechnology began while a student at MIT in the 1970's.

Nanomanufacturing is based on technologies that will affect the modern engineer by allowing him to make products smaller, stronger, less expensive, lighter and more precisely produced. The basics of nano-technology are to rearrange the building blocks of nature.

Nature provides existing proof of functioning molecular machines in the form of life. Starting with the relatively general molecular machinery common to all cells and the specific program for an individual contained in that cell's DeoxyriboNucleic Acid (DNA), one cell can build a complete organism of immense complexity. The challenge is to take control of a wider variety of materials and products than nature provides.

In the acceptance of these developments there must be a conscience of engineering. The crux of this new technology must be tempered in accordance with principles of right and just behavior. Public concern should necessitate the scientific search for truth guided by ethics. The address of nanotechnology's moral dilemmas can come in the form of visions of great promise. However, genetic engineering can go virtually unnoticed leading to a potential risk of amassment of power for those having gained the knowledge of DNA manipulation.

The final caveat to these intimations are yet to be seen. To quote Bill Joy, chief scientist and co-founder of Sun Microsystems Inc. in the United States, at a nanotechnology conference in 1989, "We can't simply do our science and not worry about these ethical issues."

With the advancement of the Atomic Force Microscope (AFM) and the Scanning Tunneling Microscope (STM) in the 1980's, nanotechnology was able to receive a better understanding which led the way for more development. Some of the first machines could merely push atoms around a surface. The next step is to be able to assemble the atoms and molecules together into a functional machine.

A True Working Robot

The Robot Institute of America (RIA) defines *robot* as: "A reprogrammable, multi-functional manipulator designed to move material, parts, tools, or specialized devices, through variable, programmable motions, for the performance of a variety of tasks."

The word *robot* comes from Czech author Karel □apek written in the 1920 play *Rossum's Universal Robots* (R.U.R.) from the base robot-, as in robota - meaning compulsory labor, robotnik - peasant owning such labor. The play depicts robots as machines with human forms in which they do the work of the people. In the play, the human-appearing robots learn enough from people to seek power for themselves at the expense of humanity. The consequence is the robots rebel and run amok, trying to obliterate the human race.

Robots perform human tasks with the science and technology of machines. Largely, humans are captivated by their artificial intelligence. They frequently execute the ability to hold and release an object, some using a very sensitive grasp. Most robots in the industrial field can be fitted with a number of end effectors and tools. A true working robot will appear more utilitarian than the fanciful or futuristic look they are given in books and movies. Science fiction, to some extent, has added to the public's concept of robotics. For example *Star Wars* offered C-3PO and R2-D2. In Stanley Kubrick and Arthur C. Clark's film, *2001: A Space Odyssey*, a highly advanced robotic system called Hal 9000 becomes deranged when the computer malfunctions. This film offered a look into what appeared to be the goal of robotics with the evidence that caution must be used in developing new technologies so that man benefits from it and is not harmed.

The high expectation of things to come is tempered by the feeling of apprehension - recalling that only a few years ago, the wild science fiction ideas of a robot have now become a reality.

Systems of automation do have certain characteristics in common:

- They are mechanical in nature.

- They can perform operations in sequence without human intervention.
- They are provided with advanced programmed instructions controlling their operation.
- Feedback is utilized in most. Signals back to the machine are part of their output, which also affects and changes the future operations. This data can come from a variety of outward directed devices, such as vision systems or tactile sensors. Controlling a robot through feedback is called closed-loop control.

One simple example of this is the clothes washing machine. By placing the unclean clothes in the machine and setting the controls, then returning later to find clothes that have been washed, rinsed and spin-dried. All of these actions happened by input from the operator, setting in motion the mechanical operations, and performing in sequence the programmed instructions. This was all accomplished without intervention from a human.

The applications of robots are many and varied. They are found in automotive plants, hazardous environments, laboratories, low light situations, medical, military, mining, printing, security, space and underwater.

Automotive Plants

- Application of robots in the automotive field is primarily found in assembly plants. As the cars move through an assembly line, the robots perform such tasks as: cutting, drilling, painting, sanding and welding. They are also able to move parts from inventory and/or storage to the assembly area.

Hazardous Environments

Robots are able to work in areas where a human may be at injured or killed. These areas include:

- Underwater Exploration - They can be used for construction and repair of structures, gathering geological samples, performing underwater mining operations, retrieving lost objects, and research of never-before-seen aquatic life in areas too dangerous or simply impossible for humans to attain.

- Bomb disarming and defusing - "Programmable Material" as it is called by inventor Joe Michael, is a robot structure which is controlled by software that can alter its shape. This allows it to begin as one shape from a safe area and, as it moves towards the bomb and comes to an obstacle, it alters its shape, negotiates the obstacle, and continues on until reaching the threat and then again changes its shape to form a barricade around the threat before beginning to disarm it.

Laboratories

- Here they are used to deliver and return samples by programming that also includes mixing in correct proportions.

<u>*Low Light Situations*</u>

- As film is processed in near-to-complete darkness, the robot can be utilized and will most likely gain future use.

<u>*Medical*</u>

- The medical field will evolve into several kinds of applications. The foremost is in prosthetics. Hands, flexible with sensitive touch, are being developed. The research to apply artificial intelligence and robotic devices is presently underway. Service sectors are some of the first areas to implement robots and benefit from this automation. One application of robotics currently being employed is the **Robot-□** (manufactured by McKesson Automated Health Care), at Anderson Area Medical Center, AnMed Health, Anderson, SC. The robot pharmacist (Robot-□) fills prescriptions, maintains quantities (overstock and out-of-stock), dispenses prescriptions ready for patient delivery, stocks, and can even determine if the packaging is erroneous. All of this happens by the robot scanning and reading bar code labels affixed to individual packages of medicines and the bar code labels with patient information. The packaging is a clear plastic bag with a bar coding label; some packages are factory labeled, quality inspected, and come pre-sealed. Some medicines are packaged at the AnMed Center. All medicines packaged on-site are quality controlled by a pharmacist in the hospital. Since installation in May, 2002 , Robot-□ has daily filled 3,000 - 4,000 prescriptions, placing its efficiency rate at 88% - 90%. The reason for Robot-□ is the accuracy of bar-code reading to remove potential errors and increase patient safety. With the installation of Robot-□, no jobs were lost and current employee job loads were eased. Also, patient costs are kept down and the system is fully accepted and favored by the professional medical world. The software and operation of Robot-□ are maintained by a pharmacy systems manager and a pharmacy director. The entire system is run on the hospital's electrical and back-up systems, and on its own stand-alone back-up system. Reliable and quality of operation is assured by having maintenance on-call twenty-four hours a day and with never more than a two hour wait time. At this time, there are only three in South Carolina. The dimensions are thirty-two feet long by seven feet wide, and it operates very quietly as it moves linearly along a bottom guide track retrieving medicines. The author personally toured the pharmacy department and viewed Robot-□ in operation. A thankful acknowledgment to Mr. Tim Hipp, Pharmacy Systems Manager, and Mrs. Pat White, Pharmacy Director, for sharing their time and knowledge.

Military

- One program is the Man Portable Robotic System (M.P.R.S.). It will be used for urban reconnaissance combat in tunnels, sewers and bunkers. It will carry cameras for detailed viewing and has the ability for two-way verbal communication with a hostile element. In this situation the robot is basically expendable. Another program using robots is the military's Unmanned Underwater Vehicles (UUV). In the March 17, 2002 issue of *Anderson Independent-Mail* [local newspaper to the author in South Carolina] an article on page E2 gives specifications for Robot Submarines (Krane, J., The Associated Press (2002, March 17). Drones lead the future of warfare: Robot submarines. Anderson Independent-Mail, p. E2. Their source: *Military Analysis Network).*

Mining

- Mining robots incorporate the characteristics of low light situations and military uses. Many robots can be controlled from the surface by one operator thereby minimizing the risks and health hazards that are inherent to mining employees.

Not all mining is done on Earth. The first lunar mission in 25 years was launched on January 7, 1998 from the Kennedy Space Center in Cape Canaveral, Florida. Lunar Prospector, a small, drum-shaped space-craft weighing only 295 kilograms (650 pounds), orbited at an altitude of 100 kilometers (62 miles) above the moon's surface. The data to be returned was information on the possible presence of water or ice in polar craters. Because of the moon's slight axis in relation to the sun, some of these regions never receive light from the sun, and the water in those areas would be frozen (Lunar Prospector [World Wide Web] Retrieved March 27, 2004 from the World Wide Web: http://www.lunar.arc.nasa.gov).

On July 31, 1998, Lunar Prospector was intentionally crashed into a shadowed crater on the Moon. The controlled crash near the south pole of the Moon was intended to loosen water vapor and this vapor could then be detected from Earth. Although this mission did not confirm the existence of water or ice, the ability to do future lunar prospecting was achieved.

Printing

- Robots in the printing industry can be compared to the laboratory setting, where they can deliver and return supplies, apply book covers, and be used for sorting and typing of materials.

Security

- The advantages here are numerous. Robots can be programmed to follow a predetermined path in an office or warehouse setting. The robot does not get tired, is always on the job, and can provide real-time asset tracking.

Space

- Many robots have been used and are presently in use in the space program. One such robot is, "Sojourner." In 1997, the Mars Pathfinder landed on the surface of Mars, kicking off one of NASA's most successful missions. The Pathfinder carried a small, six-wheeled robot, which roamed the surface of the planet and collected valuable data for researchers.

In 2003, NASA launched a larger version. This rover explorer will be able to trek up to 100 meters (328 feet) each Martian day (which is 24 hours and 37 minutes long). One of its missions will be to search for the presence or evidence of liquid water. It also carried a tool called the "RAT" or Rock Abrasion Tool; this will be used to expose fresh rock surfaces for study. Robots are also used in many ways in scientific research, particularly in the handling of radioactive or other hazardous materials. Many other highly automated systems are also often considered as robots. These include the probes that have landed on and tested the soils of the Moon, Venus and Mars, and the pilotless planes and guided missiles of the military.

<u>*Underwater*</u>

- The Earth's surface is largely covered in water. According to Webster's World Millennium DVD Encyclopedia© - The Cambridge Encyclopedia, "The earth's surface is covered by 70% water" (Water Introduction. <u>Webster's Millennium 2000 DVD Encyclopedia©</u>. 4092/42655. [CD-ROM]. Webster Pub-lishing and/or licensors). Yet the research to discover more about this mainly hidden world is slow to progress. One primary reason is that man is not suited to extreme depths underwater nor can divers stay down for extended periods of time. This is due to the fact that as the depth increases there is also an increase in hydrostatic pressure. Until technology can provide the ways and means to gain access to these areas, it will remain obscured to future discoveries. At present, a

> sport diver can reach depths of 40 meters (131 feet), and 60 meters (197 feet) for divers with special air mixtures. A whale has the capability of reaching depths of 1,000 meters (.6 mile).

Service robots will be ideally suited to going to depths that are out of reach of divers. Deep-sea exploration is currently carried out by a robot called ABE (Autonomous Benthic Explorer). ABE's maximum depth is 5,500 meters (2.8 miles) with a range of 10 to 100 kilometers (6 to 62 miles) and a speed of up to 2 knots (2.3 mph).

Robots in the future will continue to improve as technology improves. Progress with the autonomous robot awaits significant advances in other closely related devices and systems. These systems include high-capacity sensors, machine vision, speech recognition, parallel processing computers, advanced software, and dramatically improved batteries and alternative energy storage devices.

The future will also see the use of miniature robots or biobots. As with all new technology, the practical applications of small biobots are not fully understood. Some applications may include using them to detect and repair cracks in pipes or gas tanks. These new technologies are dependent upon micromachining because of the necessity for smaller gears and mechanisms to operate them.

Sensible Limits and Beyond Nanotechnology

Nanotechnology is the ultimate in manufacturing technology. It combines various criteria which make it very attractive today. One of these reasons is precision - this is accomplished with atom-by-atom placement. The gain in such precise building is the ability to manufacture complex structures and demonstrate that it is not only possible but economical.

Ralph C. Merkle, then a research scientist at Xerox Palo Alto Research Center (PARC) in California, former Principle Fellow at Zyvex and who also received Foresight Institute's 1998 Feynman Prize in Nanotechnology for Theoretical Work, is quoted as illustrating nanotechnology by a description of trees. "Trees grow by taking energy from sunlight and nutrients from the soil to build themselves. They do it quietly and discretely, without digging holes in the ground or burning fossil fuels or throwing off noxious waste products. They use only what they need, arranging the atoms in complex internal patterns. And trees also self-replicate: They produce seeds that build other trees. Precisely because it's a miracle of biology, lumber costs only a few dollars a pound." (Fouke, Janie, Editor, Trudy E. Bell and Dave Dooling, Writers. Engineering Tomorrow: Today's Technology Experts Envision the Next Century. IEEE Press Marketing. 2000)

Robotics

With the growth of technology, the use of robots continues to grow along with it. In the future, robots will have optical character recognition verification whereby a robot could read lines of print or recognize colors. Some types of robots are master-slave styles. They incorporate the use of a human wearing sensors. The motions are linked to a computer and the robot mimics the movement in speed and smoothness.

Current work on industrial robots is devoted to increasing their sensitivity to the work environment. Computer-linked television cameras serve as eyes and pressure-sensitive "skins" are being developed for manipulator grippers. Many other kinds of sensors can also be placed on robots. Paramount for robots is to have them capable of intelligent travel with a preprogrammed search (Robot - Research. Compton's Interactive Encyclopedia© 2000. [CD-ROM]. The Learning Company and its affiliates and licensors). This advancement will be dependent upon other fields in development such as advanced software, machine vision, speech recognition, and efficient miniature motors developed with the help of micromachining and nanotechnology. The fantasy of science fiction will come about with the eventual development of autonomous and service type robots.

No Boundaries for Today

A world of nano-sized machines built one atom at a time with the ability to reason and have intelligence on the level with humans. This is the future for mankind...this is the future of engineering. Throughout time man has made a forward march, not as an individual, but as a group. With discoveries being as unique as the people who create them, the discoveries show their inventor's worth is greater than the sum of the invention. As necessity is said to be the mother of all inventions, then desire should be the father. The accomplishments engineering has made to this point have been brilliant, but the progress has only scratched the surface of man's potential. The modern-day mechanical foothold can be compared with that of an inclined plane where progress has been steady and progressively upward, and one in which the past is known and all is not doomed. Only by daring to dream the impossible can the possible be realized.

Imagine a nanomachine entering a human heart to remove a blockage that otherwise may have necessitated an invasive surgical procedure. Even this is not the culmination of these technologies. What threshold is set for humans to cross? It is the limit of the imagination and beyond. How is it that all this technology will be used wisely and not for sheer exploitation? This question is yet to be answered.

If the future allows the boxing gloves to be removed, then today's stumbling blocks will become tomorrow's building blocks. The time is now to bond the confluence of the sciences and humanity for care with regard to the future should be taken as to what is built.

As a Mechanical Engineer and author-researcher, the management and mismanagement of such futuristic technologies give rise to some degree of concern. If the concerns of a country being able to gather and use this technology for terrorist intentions are unsettling, then there is a reason to be careful.

CHAPTER 2
Review of Related Literature

"There's Plenty of Room at the Bottom"

This new technology which resides on the head of a pin (and decidedly smaller), is normally referred to as nanotechnology. This technology has been sensationalized by such publications as *Popular Science* with its headline story of "How To Build An Automobile Smaller Than This Dot · ." The term nanotechnology is the larger or the broader definition; the terminology given by *Nanotechnology: Zyvex* is "molecular nanotechnology" or "molecular manufacturing" (Nanotechnology. [World Wide Web] Retrieved March 25, 2004 from the World Wide Web: http://www.zyvex.com/nano). The conditions for nanotechnology allow the following:

- Exact placement of every atom.
- Make almost any structure consistent within the lws of physics and chemistr that can bespecifediatomic detail.

- Have manufacturing costs not greatly exceeding the cost of the required raw material and energy.

These are the qualities that are necessary to continue the drive for future improvements in the lives of all, from computers to modern materials to the ultimate goal of nanotechnology, the self replicating machines. The modern engineer will be allowed to make products smaller, stronger, less expensive, lighter and more precisely produced. These are the basics of nanotechnology, but practical applications vary greatly; in some areas such as electronics and micromachining the nanoscale is readily accessed. In others, for example biology, the promise is great but so are the technical problems that are to be overcome.

Though all of this "small world" manufacturing is beginning to take shape and Dr. K. Eric Drexler is notably a senior researcher, his inspiration was directed by Dr. Richard Feynman's talk, *There's Plenty of Room at the Bottom,* at the American Physical Society at the California Institute of Technology (Caltech) on December 29, 1959.

Dr. K. Eric Drexler, on June 26, 1992, in a conversation before the Washington Senate Committee on Commerce, Science, and Transportation, Subcommittee on Science, Technology, and Space, replied to then Senator Al Gore's statement that Dr. Richard Feynman, quantum theorist and Nobel Laureate, had made a speech in 1959 outlining the whole field [nanotechnology]. Dr. Drexler replied, "Feynman did indeed point in these directions, in a talk in December of 1959, and that has been an inspiration to us all." (Regis, Ed. Nano the emerging science of technology: remaking the world - molecule by molecule. Little, Brown and Company. 1995).

To set up the current trends, two dates in history are early notable milestones in nanotechnology. In Dr. Feynman's 1959 speech two prizes were offered: one for, "...$1,000 to the first guy who can take the information on the page of a book and put it on an area 1/25,000 smaller in linear scale in such manner that it can be read by an electron microscope." and one for, "...$1,000 to the first guy who makes an operating electric motor--a rotating electric motor which can be controlled from the outside and, not counting the lead-in wires, is only 1/64 inch cube."

In 1960, the motor prize was claimed by William H. McLellan, Mechanical Engineer. Unfortunately, he used conventional mechanical techniques which was a disappointment to Dr. Feynman who was looking for new technology instead of smaller, existing technology.

Nearly sixteen years later in 1985, Thomas Newman, a Stanford University graduate student used Charles Dickens' novel, *A Tale of Two Cities,* to win the second prize by reducing the text from a page of a book to an area 1/25,000 in scale (and remaining legible) by an electron microscope.

Top-Down and Bottom-Up

From Gregory Timp in *Nanotechnology*, the current way used to build things is accomplished by a method known as the *top-down approach* (Timp, Gregory (Ed.). Nanotechnology. AIP Press - Imprint of Springer-Verlag New York, Inc. 1998). What that is simply saying is to take an object and remove the unwanted elements until the desired item is left. This approach is used by the electronics industry using lithographic techniques to produce circuits out of silicon. Most machining operations perform the same way, for example, the way a drill bit removes material to produce a hole.

The opposite is true for drug and chemical industries. These areas use the *bottom-up approach*. Through the use of thermodynamics, the chemical producer can swirl together chemical substances in quantities so that the desired molecules are combined to make a product.

Nanotechnology is achieved by combining the methods of *top-down* and *bottom-up*. This embodiment of blending the two, by building things one atom at a time as in chemistry, and by atomic scale construction as in electronics, results in the product being built are just what is wanted and within exact specification (Timp, 1985).

Three distinct nanotechnologies are described here under the guidance of Richard Smalley at Rice University. Dr. Smalley a University Professor, Gene and Norman Hackerman Professor of Chemistry and Professor of Physics & Astronomy, 1996 Nobel Prize Winner. This analysis gives the capability for understanding, fabricating, and manipulating structures at the atomic level.

Wet nanotechnology is the study of biological systems that exist primarily in a water environment. The functional nanometer-scale structures of interest here are genetic materials, membranes, enzymes and other cellular components. The success of this nanotechnology is amply demonstrated by the existence of living organisms whose form, function, and evolution are governed by the interactions of nanometer-scale structures.

Dry nanotechnology, which derives from surface science and physical chemistry, focuses on fabrication of structures in carbon (e.g. fullerenes and nanotubes), silicon, and other inorganic materials. Unlike the wet technology, dry techniques admit use of metals and semiconductors. The active conduction electrons of these materials make them too reactive to operate in a wet environment, but these same electrons provide the physical properties that make dry nanostructures promising as electronic, magnetic, and optical devices. Another objective is to develop dry structures that possess some of the same attributes of the self-assembly that the wet ones exhibit.

Computational nanotechnology permits the modeling and simulation of complex nanometer-scale structures. The predictive and analytical power of computation is critical to success in nanotechnology: nature required several hundred million years to evolve a functional wet nanotechnology; the insight provided by computation should allow reduction in the development time of a working dry nanotechnology to a few decades, and it will have a major impact on the wet side as well.

The classification in all three nanotechnologies are dependent upon each other and have overlapping interests.

In 1990 a proof-of-concept was demonstrated when IBM scientists arranged thirty-five xenon atoms to spell "IBM" on a nickle surface using the tip of a scanning probe microscope. This event dramatically demonstrated that placing atoms and molecules one at a time was not only possible, but economical as well.

Selective Stickiness, according to then Xerox's PARC, former Principal Fellow at Zyvex Ralph C. Merkle in Engineering Tomorrow: Today's Technology Experts Envision the Next Century (Fouke, 2000), is a basic principle in self-assembly. Because self-replication is a naturally occurring act of biology, the cost is very economical. The usefulness of building nanotools by selective stickiness is in the tendency for two molecular parts having complementary charge patterns and shapes to stick together and form a larger part.

Precise positioning is proposed by Merkle and others using molecular-scale devices and one-ten-millionth-scale robotic arms that could systematically add or withdraw atoms by sweeping back and forth over a surface.

To facilitate the speed at which molecules are placed into position, first an assembler is required - "a small, computer-controlled robot that can be programmed to build almost anything. Then program the assemblers to replicate themselves. Last, when you have enough assemblers, build the product." With all the assemblers working simultaneously on a product, "You could build a car in a few hours" - instead of the time required by conventional production methods. "Many hands make light work."

To conclude, Merkle replies to the question of when nanotechnology is to become reality, "I don't know, ten years, fifty years? I think it will be faster than that. There is no reason in principle we should not have at least some molecular manufacturing capabilities by 2010 or 2020. It all depends on when we start!" (Fouke, 2000). Dr. Merkle's further interests include: cryonics, medical applications of nanotechnology, computational chemistry, reversible computing, neuroscience, and extropians.

Research areas considered relevant to molecular nanotechnology and molecular manufacturing include, but are not limited to:

- Molecular Electronics
- Scanning Probe Microscopy
- Materials Science
- Natural Molecular Machines
- Artificial Molecular Machines
- Artificial Self-Replicating Systems

- Computer Science
- Mechanical Engineering and Robotics Applications
- Relevant Chemical Systems - some are: Fullerenes, Diamond, Biomolecules

Robots are Economically Viable

The numerous examples of service type robots given in *Service Robots: Products, Scenarios, Visions* indicate there are as many uses for robots as there are inventors and needs (Schraft, Rolf Dieter and Gernot Schmiere. Service Robots: Products, Scenarios, Visions. A K Peters, Ltd. 2000). The field of robotics is diverse and can fit into practically all aspects of life. However various, the uses all have a basic measure to provide. These are:

- They are dependable.
- They are economically viable.
- They are flexible and individual in use.
- They produce at a high standard.

- They greatly reduce or eliminate employee injuries.
- They are socially and ecologically acceptable.

The reliable execution of a robot and its given task depends on the accuracy of the robot, the correctness of the task description, and the certainty of its environment. Real-world uncertainties often interfere with the robot's task. Unexpected events such as misplaced or missing parts cause the robot to fail completely, no matter how minor the disturbance may be, unless the robot has the ability to detect changes in the environment through its sensors and to re-plan its task. The process of dealing with failures begins with "error detection." Once an error has been detected, the goal is "error recovery." Error recovery has been defined as the ability in intelligent robotic systems to detect a variety of errors and, through programming, take corrective action to resolve the problem and complete the desired process.

The dependability of robots, in general, relies on using highly reliable components and designs to minimize the probability that a fault will occur. In industrial robotic systems for a two-shift operation, maintenance may cost up to 10% per annum of the original purchase price.

In contrast, the Lunar Prospector had a limited life with a high price tag. On July 31, 1998, on a data gathering mission, it was intentionally crashed into a shadowed crater on the Moon. The intent was to loosen water vapor and this vapor could then be detected from Earth. This robotic mining mission was at a cost of 63 million dollars, in which 34 million dollars were for development, 25 million dollars for the launch vehicle, and 4 million dollars for operations. Ultimately the robot was destroyed but the information gained was invaluable to future space programs.

In designing robots for high reliability, there are problems caused by the fact that the environment can dictate the proper protection required. Once outside the confines of an assembly line, a robot's need for adaptability increases. The areas that robots operate in can include: heat and possibly direct flames, salt-laden atmospheres, shock and vibration, electrical noise and interferences, water (salt and fresh), dust, light or dark, and atmospheres free of gravity. Nevertheless, over the past decades of robotic use, parameters have been determined which offer guidelines for robotic design.

Service robots in the future are to become equivalent to the personal computer in the home. To date, some homes (or mansions) even have a personal robot, for example Hugh Hefner at the Playboy mansion has one to bring drinks and welcome guests.

The service robot of the future will accommodate humans in all manner of ways. Automotive refueling is one of these methods (Schraft and Schmierer, 2000). From the Bureau of Transportation Statistics: 2001 Pocket Guide to Transportation: Chapter 3: Mobility: Table 7, Number of Aircraft, Vehicles and Vessels Mode: Passenger cars 1998 indicates: number of vehicles to be 131,838,538 (2001 Pocket Guide to Transportation: Chapter 3: Mobility: [World Wide Web] Retrieved March 25 , 2004 from the World Wide Web: http://www.bts.gov/publications/ pocketguide). With this amount of traffic on the road today, if drivers were able to refuel and not exit their cars, it would make a safer and healthier environment. Robin, the automated robot-controlled refueling system has been available for state-run buses since 1993 in Saarbrücken, Germany. This method will also offer an important advantage to the handicapped (Schraft and Schmierer, 2000).

Although the term "robot" did not come into use until the 20th Century, man's fascination with robots was found in a 1773 design. French inventors Pierre and Henri~Louis Jaque~Droz, a father and son team, built an android style robot controlled by cams and driven by springs which could draw, play musical instruments, and dip a quill into an inkwell and write text up to 40 characters. According to *How The New Technology Works [a guide to high-tech concepts*] (Cone, 1998) this machine still works and can still be seen at the Musèe d'Art et Histoire in Neuchâtel, Switzerland.

"Now, look, let's start with the three fundamental Rules of Robotics."

Isaac Asimov (1920-1992), the author of more than 400 books on a broad range of subjects, called himself a "born explainer" (Robots - Isaac Asimov. Compton's Interactive

Encyclopedia© 2000. [CD-ROM]. The Learning Company and its affiliates and licensors). His streamlined versions of science facts are as popular as his science fiction, and his works include history and mysteries. He began to write and publish stories about robots in *Astounding Science Fiction* at the age of nineteen. The year was 1939 and by 1950 a collection of nine of these stories was published entitled *I, Robot*. In March 1942, a short story called "Roundaround", Published by Street and Smith Publications, one of the characters said "Now, look, let's start with the three fundamental Rules of Robotics." The three rules have become recognized as Asimov's Laws of Robotics and is the first known use of the word "robotics."

The three laws are:

1. A robot may not injure a human being, or, through inaction, allow a human being to come to harm.

2. A robot must obey the orders given it by human beings except where such orders would conflict with the First Law.

3. A robot must protect its own existence as long as such protection does not conflict with the First or Second Law.

The above laws are given in *Robots: Machines in Man's Image* (Asimov and Frenkel, 1985).

In 1985 Asimov developed this idea further by postulating a "zeroth" law that placed humanity's interests above those of any individual while retaining a high value on individual human life.

Zeroth law: A robot may not injure humanity, or, through inaction, allow humanity to come to harm.

Even with the newness of science fiction writing, the original and now continuing intent for robotics is to aide humans.

Since that time a new set of human factor rules has been proposed taking into consideration the limitations in human perceptual, decision, and action processes. These new rules are supported by Yokomizo, Y., Hasegawa, Y., and Komatsubara, A. (Graham, Ph.D., P.E., James H. Editor. Safety, Reliability, and Human Factors in Robotic Systems. Van Norstrand Reinhold, 1991.)

Rule 1: Robots must be made and used for the well-being and development of people.

Rule 2: Robots must not replace people on jobs people want to do.

Rule 3: Robots must be built to such specifications that they do not psychologically or physically oppress people.

Rule 4: Robots must follow the command of people so that they do not harm other people, only damaging themselves.

Rule 5: If robots replace people on certain jobs, the prior approval of the people affected must be obtained.

Rule 6: Robots must be made so that they can be easily operated by people and readily perform the role of assistants to people.

Rule 7: As soon as robots finish their assigned tasks, they must depart from the place so that they do not interfere with people and other robots.

In a conversation between Grant Fjermedal, longtime science and medical writer, and survey interviewer, and Harley Shaiken, research associate in the Program in Science, Technology, and Society at MIT, the content of rule # 2 and rule # 5 was discussed. Fjermedal visited the AI (Artificial

Intelligence) Lab at MIT, where he was shown a robot assembler that was capable of assembling electric drills faster than humans. Here he was told by a student that the work being displaced by the robot was "subhuman" and that, "Robots will only take the undesirable jobs." When Fjermedal questioned Shaiken as to whether he believed if workers should decide if the task being performed is "subhuman" and worthy of human participation. Shaiken replied that "...It should be up to the people to decide if they enjoy their work. Robots are going to be used where they are cost-effective. It has nothing to do with the quality of life on the job." Further Shaiken expressed his concern that those workers remaining in factories will become bracketed between robots and automated systems - becoming cogs that must operate at the same speed and with the same regularity as the automated components upstream and downstream from them.

Today one of the driving forces in most any endeavor is economics, and robotics is no different. In a March 5, 1984 interview, Lester C. Thurow (Professor of Finance and Economic Management at MIT Sloan School of Management) is quoted as saying, "No country can be fully competitive in new manufacturing industries, including robots, if it falls behind in building the tools needed to be a low-cost manufacturer" (Asimov and Frenkel, 1985).

In the end, the decision to utilize a robot in industrial or service applications is generally made on economic grounds (Susnjara, 1982). Output per hours worked is many times expressed as productivity (Asimov and Frenkel, 1985). A measure of economic strength can come from the productivity growth rate which is both a contributor to and an indication of a nation's Gross National Product (GNP). Presently the productivity in robotics is being driven by industrial robots and they are used primarily in the automotive and airplane factories. Most of the tasks assigned to robots are some type of welding, spray painting or other operations where humans would require expensive ventilating systems. This field is the most exploited and the most successful. It is done so by making robot manipulators without trying to duplicate all of the motions of which the human arm and hand are capable. The robots are also easily reprogrammed and are therefore more adaptable to changing tasks on an assembly line. The majority of industrial robots in use in the world today are moving toward factory uses and some industries even have an assembly line of robots to make still more robots. As this becomes more prevalent in the industrial field of robotics, the service sector will also gain in the use of robots (Robot - Modern Development. Compton's Interactive Encyclopedia© 2000. [CD-ROM]. The Learning Company and its affiliates and licensors).

The future economic reasons and the drive for more diverse, humane, and safe methods for workers may see the service sector as an area for increased growth. Implementing service robots on a greater scale in contrast to industrial robots is currently becoming true. According to Schraft, Rolf Dieter and Gernot Schmiere (2000), "If one sees the service robot as a halfway step in the evolution of industrial robots, then this product indeed has all the prerequisites for becoming a key product in the near future."

"One geographic area which is having difficulties is the Middle West. There is trouble on the farms, in the factories and in the cities. People are deserting the American heartland for the Sunbelt, especially Florida, Texas and California. Agribusiness continues to grow and family farms continue to diminish. The iron-and-steel and automobile industries are retooling and modernizing. They are turning to robotics and are placing new emphasis on quality and productivity. There is a trend toward diversification: computers, electronics, communications, and microbiology - the high technologies - are booming and growing, as are service industries. Retraining and new training for the job market are needed. Although great strides have been made, there is further need for revamping the region's cities" (Robots - United States the Future. Compton's Interactive Encyclopedia© 2000. [CD-ROM]. The Learning Company and its affiliates and licensors).

The future and premier objective of robotic designers and researchers is to have autonomous robots with the artificial intelligence to be self-navigating (Cone, 1998). This is currently not the case. The cost of a new robot is about $10,000 or more and, combined with human versatility, has lent itself to keeping the robot confined to jobs that are mainly dangerous, monotonous or repetitive jobs of which employees and the human mind were not intended to do.

Though these are trends based in both economics and supply and demand, they are the decisions engineers and factory management will essentially use as the current job-market indicator. Nevertheless, human activity will no doubt see the future role of autonomous robots increase and play more of an important function than today's society dictates.

CHAPTER 3
Evolutionary Stages

As with all new ideas and technologies, the direction is controlled at first by a few select pioneers and innovators, then by stages of like-minded people who are drawn to understand the potentials. Research of present-day and futuristic subjects also finds a genre of attention focused at the problematic areas of research. Presently the areas these two subjects fill are being carried out by large corporations with capital budgets and have in their employ, personnel whom are well trained.

With the uniqueness of nanotechnology and the forward movement of robotics in the service sector, research is well suited for the historical study method. In this chapter, both sides of the coin are given so as to keep the subject matter factually correct and well grounded.

Nanotechnology

Zyvex, a privately held company, is the first research and development company dedicated to developing a molecular manufacturing assembler: a device that can build materials by manipulating individual atoms. In principal, an assembler could build practically any material - diamonds, gold, and even wood. Started in April 1997, and headquartered in Richardson, Texas, the company has hired its initial researchers and is building a lab capable of the task it has set itself.

Jim Von Ehr II, founder, president and CEO of Zyvex, is moving the company forward financially by means of awards, research collaborations and from his personal finances. According to the *Dallas Fort Worth TechBiz*, Jim Von Ehr is listed as one of twelve people profiled in their "People to watch in 2002" article.

In spite of being a small company, Zyvex and its joint venture partner, Standard MEMS, Inc., in October 2001 received a $25 million, five-year, cost-shared grant under the National Institute of Standards Technology's (NIST) specifications. Both companies have also announced a two-year collaboration. In this common-goal program to produce 3-D assembled microsystems, a process is used that combines surface and bulk silicon machining in one run. The components can then be picked off the bulk silicon wafer and assembled. The actual production microsystems are expected to take another 12 to 18 months to develop.

Zyvex's first milestone was accomplished in 1999. Zyvex and Washington University in St. Louis were able to pick up single atoms using a scanning probe microscope. "Unfortunately, the atomic force microscope does not give us the precision we need," according to Ehr. Since that time Ehr has created three separate company divisions: the "bottom-up" lab, composed of molecular chemists trying to get atoms to bind to each other; the "top-down" lab, dedicated to using microelectromechanical systems (MEMS) as the basic cogs and wheels of the nano-assembler; and the "advanced systems" group, dedicated to finding the properties of individual atoms. Zyvex plans to achieve its major goal of revolutionizing manufacturing technology with emphasis on positional assembly and manufacturing in less than 10 years (Zyvex - Nanotechnology. [World Wide Web] Retrieved March 3, 2004 from the World Wide Web: http://www.zyvex.com/nano).

Nanotechnology is presently a vision rather than a reality. The research effort is in a discovery stage. Methods are drawn from chemistry, engineering science, and physics. Complex electrical circuits and materials, and machines on the 1 - 100 millimeter scale, do not exist. Nevertheless, the push to move ahead in nanotechnology and robotics is aided by organizations who offer cash prizes for those who believe that no insurmountable barriers have or will materialize.

At the Foresight Institute web site (Feynman Grand Prize [World Wide Web] Retrieved March 23, 2004 from the World Wide Web: http://www.foresight.org/grandprize.1.html), there is found a $250,000 cash prize offer by the Foresight Institute for the first person who can design and build two nanotechnology devices.

The first device is a nano-scale robotic arm, "...or other positional device that fits into a cube no larger than 100 nanometers in any dimension," which is functional and performs according to set specifications.

The second device is to construct a computing machine capable of correctly adding two 8-bit binary numbers, and all this must be nanoscale, in specifications of, "...no larger than 50 nanometers in any dimension." Both of these prizes are offered under the title of "Feynman Grand Prize" in honor of Dr. Richard P. Feynman. The Foresight Institute (a not-for-profit organization located in Palo Alto, CA) has a mission goal as related to "molecular technology" and to "...the betterment of the human condition." According to Dr. K. Erik Drexler, "The Feynman Prize will recognize one of the most significant technological breakthroughs in human history. However, the rewards awaiting those who achieve significant nanotechnology breakthroughs will be far greater than the prize itself."

Although much of nanotechnology is in the beginning stages of development or altogether non-existent, there are some who debate the possibilities of nanotechnology coming to fruition.

The September 2001 issue of *Scientific American* contained an article by Nobelist Richard E. Smalley (A Debate About Assemblers [World Wide Web] Retrieved March 22, 2004 from the World Wide Web: http://www.imm. org/sciamdebate2/ smalley.html). In the article he stated, "Self-replicating, mechanical nanobots are not simply possible in our world." He also stated two primary objections in support of his assertion: "fat fingers" and "sticky fingers," "and neither can be avoided."

The "Fat Fingers" Problem

The importance of Smalley's argument is in his statement of the number of reactants in this. He says, "There just isn't enough room in the nano-size reaction region to accommodate all of the fingers of all the manipulators necessary to have complete control of the chemistry."

The "Sticky Fingers" Problem

In this area Smalley claims, "...the atoms of the manipulator hands will adhere to the atom that is being moved. So it will often be impossible to release this minuscule building block in precisely the right spot...these problems are fundamental...."

These problems should be noted, carry objections from Smalley, and refer to designs he himself proposed (http://www.imm.org/ sciamdebate2/smalley.html) "not the designs of Drexler, whom he attempted to refute."

At the web site, both problems were rebutted with experimental evidence by Wilson Ho and Hyojune Lee.

Robotics - From a Small Lizard To Deep Reasoning

The validity of robotics is one area in which numerous institutions are and have been involved. One recognized institution is The Robotics Institute at Carnegie Mellon University. It was established in 1979 to conduct basic and applied research in robotics technologies relevant to industrial and societal tasks. While retaining its original goal of realizing the potential of the robotics field, the Robotics Institute has diversified its efforts and approaches to robotics sciences.

The Robotics Institute has a comprised funding totaling $24 million and has received funding from: NASA, U.S. Air Force and U.S. Army, among other federal, industrial and non-federal sponsors and affiliates.

A principal research scientist at Carnegie Mellon University Robotics Institute is Hans Moravec. According to Hans Moravec in an interview given October 1997 with NOVA (Robots - Hans Moravec [World Wide Web] Retrieved March 25, 2004 from the World Wide Web: http://www.pbs.org/ wgbh/nova/ robots/moravec.html), the working definition of a robot is, "...machines that are able to do things that previously had been associated only with human beings or animals – such as the ability to understand their surroundings and plan their actions." He also replied to NOVA's question about an earlier interview in which Moravec said that, "Today's best robots can think at an insect level." Since that time and at the time of NOVA's interview [1997], Moravec has updated his response to, "....So we're at the stage of small vertebrates. Over the next few decades, the power [of computers] is going to take us through small mammals and large mammals." Moravec is suggesting that by the middle of this century, the techniques on robotics will get to the human level. "This will happen in four evolutionary stages" stated Moravec.

The first stage will be a small machine that is able to maintain a very dense map or image of its surroundings. It will be able to both plan its actions and to navigate so that it knows at every moment where it is. Moravec thinks that some time within the next five to ten years there will be something like that. (Time line about the year 2010 and comparable to a small lizard.)

Second stage robots will be more flexible and reliable. This stage will see robots with motion planning and collision avoidance. This will allow them to "learn" from their mistakes, an improvement over stage one robots. (Time line about the year 2020 and comparable to a small mammal.)

Third stage robots will be able to think before acting. This is accomplished by a module allowing the robot to make mistakes in its mind, running through scenarios in simulation rather than physically. (No time line given, and comparable to a monkey.)

Fourth stage robots basically will marry the third stage robot's ability to simulate the world with an extremely powerful reasoning program. An example in reasoning power is Deep Blue, IBM's supercomputer made famous by its 1997 chess victory over Grandmaster Garry Kasparov.

By coupling reasoning power to physical experience or understanding, situations that the robot thinks about in its simulation can now be abstracted into statements about the world. (Time line about the year 2040, and probably more deep reasoning power than humans.) This interview is accessible by going to http://www.pbs.org/wgbh/nova/robots/ moravec.html.

Another interesting point brought up in the interview is the ability of a robot to deduce the state of humans around them. Massachusetts Institute of Technology (MIT) is presently doing research in "mood recognition" at their Media Lab. These robots should be able to deduce if a human seems to be in a hurry or seems to be tired.

Another project currently on-going at MIT is Kismet, the robot. At the MIT web site (Kismet, the robot. [World Wide

Web] Retrieved March 27, 2004 from the World Wide Web: http://www.ai.mit.edu/projects/ sociable/baby-bits.html) Kismet can be found. Kismet is designed to be altricial in nature. The aim is for a benevolent "caretaker" to interact similar to a mother and child. This research is directed at how to build a more open-ended learning system. Kismet has expressive capabilities to facilitate a more natural infant caretaker interaction. The robot has vision, auditory, and proprioceptive sensory inputs. The course for Kismet is to apply many of the same social indications that human infants are exposed to as they learn to interact with their parent or "caretaker."

Nova, Carnegie Mellon and MIT

In both fields, nanotechnology and robotics, only primary sources were chosen for discussion of topics. Representative for the robotics approach in this chapter is NOVA's interview with Hans Moravec, a Principal Research Scientist at Carnegie Mellon University Robotics Institute.

NOVA is known for their contribution to the scientific world and the proficiency in showing the human story behind the science story. They also give Internet audiences real-time travel with explorers and scientists to all parts of the world.

In addition to winning every major television award, NOVA recently won an Emmy, a DuPont Columbia, and a Peabody award. With NOVA's programming and first-hand interviews it will continue to be a great source and resource for any scientific and selected themes of interests.

Massachusetts Institute of Technology (MIT) is another source referred to in this book. MIT was chosen because of the length of time involved with robotics technology and with the numerous other academic disciplines that are clearly geared toward developing the field of robotics. Over 200 people are actively involved in the many laboratories and departments associated with robotics. Some of the areas MIT is pursuing are:

- Bio Machines
- Humanoid Robotics

- Machine Learning
- Medical Vision
- Speech
- Vision

Peripheral Systems
A Blueprint For Nanotechnology

Many researchers wish to adopt a definition of "nanotechnology" which includes their own work (Foresight Debate with Scientific American; Foresight Institute vs. Scientific American; Debate on Nanotechnology: Round 1 from Merkle. [World Wide Web]. Retrieved March 22, 2004 from the World Wide Web: http://www.foresight.org/sciamdebate/sciam response.html). An unfortunate consequence of this is that the unqualified term "nanotechnology" has come to mean very little. In the present critique "nanotechnology" is defined as a synonym for "molecular manufacturing." In contexts where it might be unclear which usage is appropriate, the term "molecular nanotechnology" or "molecular manufacturing" can be used. In any event, it is essential to establish an appropriate definition if a discussion of "nanotechnology" is to make any sense.

The limitations in nanotechnology are due to peripheral systems that nanotechnology needs so the research can progress. Some of these areas include:

- The development of nano-sized robot arms capable of manipulating atoms. Note: it took over a decade to create the first working nano-tech robotic arm.
- Continued advancements in the field of chip manufacture. The race to manufacture faster chips and realize high profits is highly competitive.
- Biologists are focusing research in genetics. This relates to nanotechnology as genetics is a blueprint in self-replicating biological life.
- To summarize: from Dr. Richard Feynman's 1959 talk, "The principals of physics, as far as I can see, do not speak against the possibility of maneuvering things atom by atom. It is not an

attempt to violate any laws; it is something, in principal, that can be done; but in practice, it has not been done because we are too big."

Once these problems are addressed or at least some of these areas are developed, the products of nanotechnology will be implemented in wide and varied fields of use.

- Moore's Law - the ability of computer chip manufacturing to double every year may slow down. Computer chip manufacturers are running into the problem of reducing chip size while increasing speed. An observation made in 1965 by Gordon Moore, co-founder of Intel, stated that the number of transistors per square inch on integrated circuits has doubled every year since the integrated circuit was invented. The current definition of Moore's Law is revised to read - the number of transistors per square inch on integrated circuits has doubled every 18 months since the integrated circuit was invented. According to

Moore, "Some time in the next several years we get to some finite limits, but not before we get through five generations." Some predictions have the physical limits set to be reached by 2017. This problem will continue until nanotechnology develops further, allowing more and more transistors on a chip, which results in an increase in performance since new performance can be added. Nanotechnology will allow chip manufacturing on an atom by atom process, permitting chips to be thousands of times smaller and millions of times faster.

- Medicine has intentions of using nanotechnology - imagine having bones wrapped with invisible diamond fabric; increasing body tolerance to "G" forces. Even more remarkable are the nanocomputers and manipulators. They will be small enough to insert into the blood stream, without compromising cellular function. One proposed function for them is to take an inventory of the host cell's structures using the cell's DNA as a blueprint. If a foreign material or cell should arrive and be found outside the original's DNA inventory, the manipulator would destroy the invader before it had time to cause damage. Former astronaut and United States Senator, John Glenn, on his second mission into space, swallowed a pill containing a miniaturized radio communications device allowing his vital signs to be monitored.

- Super materials of atomic precision - bearings devoid of defects by being produced precisely round. They would last longer, run cooler and bear greater loads. Buckytubes are the strongest material that can be produced with known matter (not considering PBO, Polybenzbisoxazole graphite epoxy, a high tensile-strength fiber and some theoretical possibilities of carbon nitrate). The strength of a buckytube is predicted to be somewhere between 1.2 and 2 times that of a diamond fiber or 100 - 150 times strong as steel at less than one fourth the weight.

- Wrinkle-resistant and virtually stain proof - the Lee Jeans Company in Merriam, Kansas is introducing nanotechnology-enhanced khaki pants. The line announced in 2002 are men's and women's pants that are advertised as performance khakis specially treated with NANO-CARE™ technology. The fabric treatment is not a coating and will not wash off; on the downside, moisture resistance tends to lock in perspiration. The enhancements are molecular-level changes made to the fabric and can only be seen with a powerful microscope making it difficult to readily present consumers with the difference between their products and their competitors. The enhancements are developed by Nano-Tex LLC, Limited Liability Company, a unit of the fabric company Burlington Industries.

- Military nanotechnology in a new era of strategic competition - the military is noticeably taking a part in this new technology. Materials of super strength and reduced weight will be of great importance to airplanes and in crafts used in the space program. Other possibilities are bullets that can pierce almost any armor. Communications will be improved by reducing the weight and size of radios and necessary equipment carried by the soldier. Real-time global person-to-person communications through personal communicators - this technology practically exists today. One factor taking high priority is the issue of national security. It is believed that the nation who can develop nanotechnology and robotics and put them into practice will become a world military power. Molecular manufacturing has even greater potential than nuclear weapons to radically change the balance of power.

Robots On A Level With Humans

Paramount for robots is to have them capable of intelligent travel with a preprogrammed search. This advancement will be dependent upon other fields in development such as:

- Advanced Software - On a futuristic note, further advancement awaits computers or machines that can process and reason similar to the capabilities of human beings.
- Machine Vision - A future robot will need to be able to "see" and recognize objects for collision avoidance and to become truly autonomous. Vision systems that let a robot adjust its behavior to meet a changing environment are among the most complex and difficult to create.
- Speech Recognition - Is a two-way street for more human-robot interaction. As this progresses,

the robot should be able to "understand" and "communicate" on a level with humans.

- Interaction - There needs to be a natural, real-time dynamic of interaction. This is a current issue that is being addressed as recounted earlier with the robot Kismet at MIT.
- Efficient miniature motors developed with the help of micromachining and nanotechnology. In addition, micro-machining and nanotechnology will become essential in being able to build smaller "biobots."

With new breakthroughs and increased volume will come greater capabilities and lower costs. This in-turn could bring about the fantasy of science fiction and the eventual development of autonomous and service type robots (Susnjara, 1982).

A Shotgun Marriage

No other field in history has necessitated knowledge from such a wide spectrum of disciplines as does molecular nanotechnology. The prospective nanotechnologist must spend more time than usual in the study of many fields, however, the potential payoff is considerable.

The future nanotechnologist should have knowledge and training in the following fields, including but not limited to: applied mathematics, art, biochemistry, biosciences, biotechnology, cellular biology, chemistry, computational biology, computer simulation/design/modeling/science, drafting, electronics, encryption, engineering, genetics, metallurgy material science, metallurgy, physics, and protein engineering. Nanotechnology is basically a shotgun marriage of chemistry and mechanical engineering, with physics invariably presiding.

The University of Washington is one of the first universities to offer a doctoral degree program in nanotechnology. Since its inception in 1997, the Center for Nanotechnology has established a goal of excellence in teaching, research, and public outreach. The prime goal of the Center for Nanotechnology is the instruction and training of future scientists and educators who will apply the knowledge and skills of this and other disciplines in their chosen fields.

The Center, directed by Professors David Castner and Francois Baneyx, was established by a University of Washingtons Initiatives Fund (UIF) award in 1997. More than fifty-five faculty members come from the Departments of Chemistry, Physics, Bioengineering, Chemical Engineering, Electrical Engineering, Materials Science and Engineering, Biochemistry, Genome Sciences, Physiology and Biophysics, and Microbiology, and is funded from the National Science Foundation's Integrative Graduate Education and Research Traineeship (NSF-IGERT) program. University of Washington's optional Ph.D. Program in Nanotechnology, the first of its kind in the nation, is providing graduate students with excellent interdisciplinary education experiences in nanoscale science and nanotechnology.

In April of 2001 the University of Washington and the Department of Energy's Pacific Northwest National

Laboratory formed the Joint Institute for Nanoscience and Nanotechnology (JIN). The activities conducted will be located on the principal campus of the University in Seattle, Washington, and Pacific Northwest National Laboratory (PNNL), as the Board and Steering committee deem appropriate. The goals and objectives, of this joint institute for Nanoscience, include some of the following: coordinate educational and training opportunities in Nanoscience for university undergraduate students, graduate students, post-doctoral fellows, and others; facilitate multi-institutional research efforts in Nanoscience, with special emphasis on partnerships among university, government, and industrial laboratories; identify additional funding from public and private sector sources for the activities of the Joint Institute for Nanoscience.

Cornell University's Nanofabrication Facility has been around for more than 25 years, offering nanofabrication services to research projects. The center has strongly developed areas such as integrated optics and microelectromechanical systems (MEMS).

The impending approach of the molecular level has stimulated a wave of interest by electronic research labs. Mitre Corporation's web site devoted to this speciality (http://www.mitre.org) reveals extensive research efforts. Other research centers listed on Mitre's pages include Massachusetts Institute of Technology, Notre Dame's Microelectronics Lab, Purdue's Nanoscale Physics laboratory and Stanford's Center has developed leading-edge expertise and equipment for ultra small device fabrication.

While current methods are just breaking ground with the individual manipulation of atoms and molecules, the full visionary idea fueling nanotechnology enthusiasts is the possibility of building manufacturing machines and robots on the nanometer scale. Like life forms, which are self- reproducing molecular machines, these hypothetical machines would be able to rapidly build billions of copies of themselves and go to work tirelessly performing the tasks they were programmed for.

Engineers would also will be able to transcend the laws of nature - or appear to. By designing different actuators on different sides of nano-sized building blocks this would make it possible to create materials with seemingly impossible properties, such as a material that never flexes all the way to its breaking strength. On a microscopic level, the real flexing could be compensated for by extending different actuators on different parts of the material's building blocks, providing the appearance of zero flex at the macroscopic level.

All of these procedures, sciences, and technologies are dependent upon each other. The future of nanotechnology and robotics will continue progressively as long as there are people trained in engineering and the sciences. As to engineer is human, the capabilities of nanotechnology and robotics in the manufacture of smaller machines, and robots, discoveries and devices are also created that before now would have been impossible.

CHAPTER 4
Pandora's Boxes

"As we enter the third millennium, I can't imagine a more important technological initiative to undertake than the National Nanotechnology Initiative (NNI). The results forthcoming from such an initiative will transform our lives and transform the very concept of manufacturing in ways that it is hard to fathom at this moment in time. The bringing together of atoms and bits raises many provocative technological and scientific questions. I believe that such a long-term initiative will have short, medium, and long term impacts and will help stretch the national imagination.", John Seely Brown, former Chief Scientist of Xerox Corporation and former Director of Xerox Palo Alto Research Center, in a National Nanotechnology Initiative (NNI) Endorsement. He currently serves on the boards of directors for Corning, Varian Medical Systems and Polycom and on numerous advisory boards and boards of startups. (World Wide Web,

Retrieved June 26, 2004 from the World Wide Web: http://www.johnseelybrown.com/bio.html)

Bill Joy, former Chief Scientist and Co-Founder of Sun Microsystems Inc. in the United States, issued this admonition titled "Why the future doesn't need us." in *Wired* magazine, April 2000. "I have always, rather, had a strong belief in the value of the scientific search for truth and in the ability of great engineering to bring material progress." Joy said he is simply calling for a return to common sense. "Reason taught us how to develop these tools and if we don't use our reason to manage them, we do so at our own risk."

The White Plague is a novel by Frank Herbert in which a molecular biologist is driven insane by the senseless murder of his family. To seek revenge he constructs and disseminates a new and highly contagious plague that kills widely but selectively. Joy asks, "Why weren't other people more concerned about these nightmarish scenarios? Part of the answer certainly lies in our attitude toward the new - in our bias toward instant familiarity and unquestioning acceptance. Accustomed to living with almost routine scientific breakthroughs, we have yet to come to terms with the fact that the most compelling 21st-century technologies - robotics, genetic engineering, and nanotechnology - pose a different threat than the technologies that have come before. Specifically, robots, engineered organisms, and nanobots share a dangerous amplifying factor: they can self-replicate. A bomb is blown up only once - but one bot can become many, and quickly get out of control. The new Pandora's boxes of genetics, nanotechnology and robotics are almost open, yet we seem to have hardly noticed. Ideas cannot be put back into the box. The fail-safe point is rapidly approaching."

Erik Drexler is quoted as saying (Regis, Ed. Nano the emerging science of technology: remaking the world - molecule by molecule. Little, Brown and Company. 1995) "I predict that the last refuge of critics will be to deny that anything short of a physical demonstration can provide solid evidence for the feasibility of something new like molecular manufacturing. This position, of course, denies the usefulness of all but the most routine sorts of engineering, and comes close to denying the value of any kind of scientific theory.

It can be called the "wake me up when it's over school of criticism."

Technology Backed by Presidents

In order to validate and confirm the findings and analyses reported, only knowledgeable and reputable universities are cited; doctors of physics, nanotechnology, and molecular nanotechnology are quoted; experts in the field of robots and robotics gave learned lectures and numerous books are referenced.

To further bolster the cause for future development the Presidents of the United States from the present and last administrations (President William Jefferson Clinton, 42nd President of the United States, two terms of office, 1993-1997, 1997-2001 and President George W. Bush 43rd President of the United States, 2001-) have both exhibited their support with federal funds set-aside for nanotechnology and robotics, among other modern technologies.

Who Controls the Controller?

Can materials and machines be constructed on an atom-by-atom basis? At the present time, there is laboratory research that suggests the answer is yes. It is important to remember that these initial steps deal with a few hundred atoms at most. So very small pieces can be made now, and presumably in a few years something will be made that can be seen with the naked eye.

To bring this technology from these initial small steps to total or partial automation, certain drawbacks and merits exist. The utilization of nanotechnology and robotics in future manufacturing will be a driving force for the trend in factory automation. To help understand this, it is essential to recognize the difference between automation and mechanization.

Automation is derived from automatic and operation. Automation is sometimes confused with mechanization, but the two are quite distinct. Mechanization involves the use of machines instead of human physical effort to accomplish a given task. True automation, by contrast, is the performance of an operation by automatically controlled equipment capable of self-regulation. Application of this new technology to itself and to automation bring up the following four basic points:

1) Although most automation known in the mechanical world today is very reliable, and following on the issue of reliability, what is to be expected of a nanosystem malfunction? From Jaron Lanier, in an article entitled, "One-Half of a Manifesto: Why stupid software will save the future from neo-Darwinian machines," published in *Wired* magazine, December 2000 issue: "Another riposte to Joy's [Bill Joy] vision is that the biotechnology industry is an incredibly fractured one, making it more difficult for one giant electronic brain to manipulate such power. And there is little difference between machine autonomy and the abdication of human responsibility. The only way machines can achieve autonomy is if humans allow them. But even that will be hindered for years by the poverty of decent software. Far from envisaging a world of super nanobots,

it may be more likely that we have a world of help-desks solving our ever-multiplying software problems."

2) Morever, an entire system may fail to operate properly if there is a single error in setting it up. Present-day programming bugs or the willful attempt of misguided persons, have been known to wreak havoc on computer systems and automated devices. There is also the need to regulate the controlling power of such immense and life changing technologies. This can also be presented with the thought of, who controls the controller?

In Lecce, Italy, January 31 - February 1, 2002, an EC-NSF (European Cooperation - National Science Foundation) Workshop on "Nanotechnology: Revolutionary Opportunities and Social Implications" was held. Under the program heading of Social and Ethical issues, the following question was raised, "Do you expect any "separation" in the society between those who operate (or even understand nanotechnology) and those who do not?"

Scientists have been described as a tribal culture that is largely unknown to the public. Surveys show public interest in science and technology, and a substantially optimistic view of them, but with a broad degree of variation. Nanotechnology will generate positive and negative effects - new industrial opportunities, but also the decline and even extinction of some industries. More knowledge of nature will result, but also a possible "nanodivide" may occur, a distinction between those who understand the new technological approach and those who do not. There is not one public but many, with various expectations, feelings and priorities. There is a call for a model of science and technology whereby all voices should be heard.

3) A backup system has to be provided or a human "override" capability built in. System operations must be set up so that they can be handled manually. Automation makes possible the performance of tasks that are well beyond the limits of human capabilities. For example, the launching of the United States space shuttle. The tracking and control requires so many complex computations with such rapid control responses that it can only be accomplished through the employment of high-speed computerized systems. Yet,

this high-tech system has a backup system and manual control ability.

4) Automated systems lack the flexibility of humans. Computer programs can be changed with relative ease although significant changes in their function may require redesigning of part or all of the equipment. Control components, action, and sensing still have to be tailored to specific applications. As a company becomes increasingly larger and more automated, larger amounts of money are required.

With nanotechnology, the ability to control, program or reprogram a nanobot will, at least at first, lay with a select few. Present-day automated systems are lacking in flexibility; nanotechnology will also remain inflexible until the knowledge to control it is acquired.

In the development of nanotechnology and robotics, particularly the service and autonomous areas, the limitations, possible omissions, possible errors, and the analysis of reliability can be found in the unseasoned state of the technology.

Technology Not Currently Bookshelf Available

One limiting factor, in particular to nanotechnology, is the lack of availability to materials for reading and reference, not including the World Wide Web. Not all libraries and schools, at present, carry books on the subject. The readiness of information on nanotechnology is not currently bookshelf available. While this is certainly true for most areas not all locations have this problem. Many major universities will solidify plans for a nanotech department - some as quickly as this year [2004].

Officials from four universities – The University of Texas at Austin, Rice University in Houston, The University of Texas at Dallas and The University of Texas at Arlington – founded an organization known as the Strategic Partnership for Research in Nanotechnology and are beginning to offer classes in nanotechnology. These four Texas universities have formed a partnership to help position the state as a center for education, research and development in the cutting-edge science of nanotechnology. The college of Austin,

Texas. (Nanotechnology. [World Wide Web] Retrieved April 29, 2004 from the World Wide Web: http://www.texasnano.com/news.html.)

Evaluation of the World Wide Web

To write about and to keep current with the pace of world wide events the World Wide Web must be included. There are also factors in using the "Web" which govern the fact gathering. Although there is no lack of information found on the Web, the data gathered from this valuable resource must be used correctly.

The information age shared by everyone today has brought with it unlimited amounts of data. This in itself is also one of its drawbacks. If someone with a computer and access to the World Wide Web creates a web site they are also able of publishing anything, credible or not, on it. Here are some guidelines to help decide if the web page is satisfactory.

According to Nell W. Meriwether (Meriwether Nell W., Successful Research Papers in 12 Easy Steps. National Textbook Company Publishing Group. 2000.), the first thing that should be done is to look for the web address. This is done to check the source. In the web address or URL find the last few letters. This is called the domain name (location of the account on the Internet). For example in the address of http://www.kw.edu, the domain name *.edu* indicates the information is from a university and should be educational. Other domains are *.com* - commercial, *.gov* - government, *.mil* - military, *.net* - network, *.org* - organization (sometimes nonprofit).

Next the reputation of the author should be determined. The educational background and experience the author holds, what credentials can be supplied, and how recent is the material all have to be evaluated. This information can usually be found in the body of the work by looking for the date of the views and the date of publication.

Then read through the article for references and determine if those references are credible. For example an engineering article could be considered credible if authored by engineers and followed by a list of references giving engineers who have worked in the same field as the article.

Determine if the information is primary or secondary. Primary sources are firsthand (from the original source). They are obtained by consulting letters, speeches, people's journals, government documents, and by conducting interviews and surveys. Secondary sources are analytical works that comment on and interpret other works, such as primary sources. Secondary sources are discussions, analysis of works or events, biographies, critical studies and reviews.

Check the objectivity of the information, its perspectives and biases, beware of advertising in the guise of information, and beware of slick Web sites presenting only their opinions.

Lastly, check the reputation of the publisher. The work should have gone through an editing process prior to being publishing by a reputable source that is academically reliable and accepted.

Note that material on the Internet is generally not refereed or monitored in any way. The reader needs to establish careful criteria to evaluate any information found there.

A Peer-Reviewed Discovery

Possible errors, omissions and limitations can sometimes be exposed in a comprehensive writing.

One such article is found in Red Herring 100: Nanotechnology [World Wide Web] Retrieved June 6, 2004 from the World Wide Web: http://www.redherring.com/insider/2002/0524/tech-rh100nano052402.html. This article is number nine in a series of daily postings featuring concise profiles of this year's top technology companies by sector. It's short, well written and articulate, and is quoted here completely. By Stephen Herrera, May 24, 2002: "A case could be made that nanotech is just so much peer-reviewed discovery and media hype in search of a commercial application and a sucker. By rights it doesn't even deserve to be called an industry. Yet nanotech is too powerful and too broadly applicable to ignore in the medical, materials, and electronics fields. This year, governments worldwide will spend $2 billion on nanotech research; 150 VC's will invest at least a few hundred million dollars in nanotech companies; hundred of startups will be created; and numerous major universities will solidify plans for a nanotech department. Some estimates put the number of firms at least experimenting with nanotechnology at 450 worldwide. There's so much interest in nanotech because the novel electrical, chemical, and physical properties that occur only on the nano scale (ten stacked hydrogen atoms equal 1 nanometer) will fundamentally change the way we innovate. Tim Harper, a nanotech expert and former engineer with the European Space Agency, says working with clusters of atoms and programmed molecules "will not just affect the things we consider advanced, it will also impact areas we consider mundane." Nanotechnology is already starting to change the way products are designed, manufactured, and refined. We think for the better."

Security of This Small-World Technology

One especially prevalent fact controlling omissions and limitations in nanotechnology and robotics is national security and the tight control of company trade secrets. Competition for the future manufacture of small inventions is fierce and the profit is immense. Biologists are focusing on research in genetics and in self-replicating biological life. Military applications are wide reaching with global consciences for the country that can harness and control this new technology. Until the breakthrough to known manufacturing methods is perfected or at least some aspect of the technology in use, the control of this small-world technology is assuredly being guarded closely.

This restrictive environment necessitates that care must be taken in future guidelines so that research and development are not driven underground. This could expose compliant countries to the increased risks associated with decreased technical, economic, and military capabilities. It would also sacrifice the many significant economic, environmental, and medical benefits that counteract serious and certain risks that society now faces in industrialized countries, and particularly in the developing world. Another goal should be to educate researchers about the potential benefits and risks of the technology. The long-term goal should be to produce a dialogue and set of guidelines that would be useful to policy makers, the public, and to the research and development community.

CHAPTER 5
The Essential Element

The approach of nanofactory automation is coming and the intrinsic nature of its introduction into individual lives dictates the methods by which nanotechnology and robotics will be used. Bank automatic teller machines (ATM), as an example, could be considered the first white-collar robots. Even with human clerks inside the bank, ATM's are now commonplace, and even expected, as a banking service.

The formulation of effective approaches to encouraging fuller realization of the constructive potentials offered by programmable automation systems requires careful consideration. Companies and management of these companies must recognize this in order to achieve improved productivity. This is being introduced through nanotechnology and robotics and other flexible machinery; basically the design of the production must be enhanced. To accomplish this there must be improved capital equipment, an emphasis on the structure and operation of the production system, and by enhanced knowledge and skills of labor and technical personnel.

During 1980 - 1990, the percentage in the age group 18 - 24 declined by 15%. As a result of this drop in new workers entering the workforce, there was a decline in unskilled labor. This created the opportunity for more automated manufacturing. Additionally, the effects of nanotechnology and robotics will be to lower cost, improve performance, and substantially widen the availability of automation technology.

In the past, despite some initial displacement of labor, labor-saving techniques have ultimately led to improved living standards and higher wages. Since total employment is a function of real economic growth, nanotechnology and robotics can have a positive effect on total employment. Employment in a given industry may rise if productivity improvements are combined with increases in production volume. In addition, average wage levels will vary with changes in the necessary mix of workers' skills resulting from their implementation. This can also benefit U.S. employment competitiveness by overcoming low wage rates overseas through a combination of automated manufacturing equipment and domestic labor. This will result in yielding a total cost-per-unit competitive with overseas costs.

How much benefit will be derived from automation in the fields of nanotechnology and robotics? How will these benefits be distributed? These questions should be addressed in the early stages of their development. These questions have a direct dependency upon how employers handle the profits - will these profits be kept or will production be expanded or will the profit be shared with workers? Worker cooperation or resistance to automation will directly affect the pace of the U.S. to compete in world markets. One method to implement that may keep from displacing U.S. employees is to make long-term investments in workers and capital investments alike.

New technology cannot serve as a substitute for good management and human motivation. Many recognize that it is the relationship of new technology and worker motivation that leads to productivity growth. The advantage of new technologies should not be seen as a disadvantage of the worker. Employees should not fear loss of jobs and the devaluation of their skills, or having to accept lower wages or increased output to prevent the loss of jobs due to the use of new technologies. There is historic evidence that automation has led to overall increased employment; nevertheless, the introduction of automation initially led to significant labor displacement. The United States is fortunate that it has industries which have the potential to innovate and economically develop technologies that has such far reaching goals. Management and labor will find new opportunities being offered in using nanotechnology and robotics.

Automated programming is usually not feasible for operations requiring complex tasks and is not routinely subject to schematized repetition. The use of new technologies in automation changes the quantity and quality of more technical operations, but this is not due to direct substitutions for a human worker.

Human and Robot: A Characteristic Look

ACTION AND MANIPULATION

Human	Robot
Two arms and two legs; multipurpose hands.	One or more arms; automatic hand change is possible.
Two hands cannot operate independently.	Incremental usefulness per each additional arm can be relatively higher than in humans.
Movement time and accuracy governed by Fitt's Law. High-precision movement may interfere with calculation processes.	

BRAIN AND CONTROL

Human	Robot
Slow - 5 bits/sec.	Fast, e.g., up to 10,000 bits/sec. for a small minicomputer control.
Affected by meaning and connotation of signals.	Not affected by meaning and connotation of signals.
	No valuation of quality of information unless provided by program.
Good error detection/ correction at cost of redundancy.	Error detection depends on program.
	Very good computational and algorithmic capability by computer.
Time lags increased, 1 - 3 sec.	Negligible time lags.
Limited ability to accept information (10 - 20 bits/sec).	Ability to accept information is very high, limited only by the channel rate.
Very limited response selection; execution (1 sec) responses may be "grouped" with practice.	Good ability to select and execute responses.

MEMORY

Human	Robot
No indication of capacity limitation.	Memory capacity from 20 commands to 2000 commands and can be extended by secondary memory.
	Can forget completely but only on command.
Memory contains basic skills accumulated by experience.	"Skills" must be specified in programs.
Slow storage access/retrieval.	
Very limited working register (5 times).	

TRAINING

Human	Robot
Requires human teacher.	Requires training by an experienced human or machine.
Usually individualized is best.	Training does not have to be individualized.
Retraining often needed due to forgetting.	No need to retrain once program is taught correctly.

SOCIAL

Human	Robot
Needs considerable support.	No social and psychological needs.

This comparison of Human vs. Robot Skills and Characteristics is referenced from Hunt, V. Daniel. Smart Robots: A Handbook of Intelligent Robotic Systems. Chapman and Hall. 1985.

As systems change, computing power becomes paramount. Description of an ideal system: Starts out weighing eight pounds and costing two thousand dollars. Its life expectancy is about sixty-five years and it takes about twenty years to bring to full programming. It has about four terabytes of available RAM and ROM. Two to three thousand dollars a year will keep the software upgraded. After that, it is essentially self-learning. It has advanced mobility, speech synthesis, stereo discrimination, voice recognition, and visual discrimination. The final system gives a return on investment of about $80,000 - $100,000. Power consumption is about one hundred watts. The final system size would be anywhere from five to six feet tall and weigh from one to two hundred pounds. (Dr. Kerry Joels, presenting a paper on his vision of future robots, at a meeting of the Association of Computer Machinery) Fjermedal, Grant. The Tomorrow Makers: a brave new world of living-brain machines. Macmillan Publishing Company. 1986.

If this description sounds similar to a person's characteristics, it is because it is a human's delineation.

People are what makes everything work. We all cultivate the needs of the future just by our natural curiosity to improve the world in which we live. Life and the pursuit of happiness make humans the essential element and by far the best all-around computer and robot (per se), and is nanotechnology at its finest.

Of Molecules and Machines

Technology and capitalism have always been partners of necessity to bring about remarkable changes. Today's current industrial revolution may be the Internet, but like nanotechnology, robotics, and other scientific advancements, the Internet has swept onto the scene with such speed and force that ethical and privacy issues have been left behind. These issues may be resolved eventually, but when nanotechnology and robotics become a way of life, these same issues are likely to continue to surface in new and varied forms. Those who take bold steps and move quickly to adopt new technologies before they become ordinary will be the trailblazers of the 21st century. Max Planck, the German theoretical physicist who abandoned classical Newtonian principals to introduce the Quantum Theory, said, "a new scientific truth does not triumph by convincing its opponents and making them see the light, but rather because its opponents eventually die out, and a new generation grows up familiar with it."

No one has advanced any credible reason for believing the approaches in nanosystems won't work, and it's very difficult to say how long it will take to develop without actually trying. A new generation of more open-minded engineers and scientists are likely to make advances in areas where their conservative colleagues are unwilling to look. These feats of engineering will ultimately become what is now posted as a touchstone, whereas the future inhabitants of this world will benchmark the designs of today as a reflection of themselves.

Nanotechnology and robotics are the history of future effects to betide.

GLOSSARY

Advanced Software: The programs, routines, or supporting documentation that instruct the operations of a computer. At present, there is software that performs sophisticated functions but is not "thinking" or demonstrating originality in any fundamental sense. On a futuristic note, further advancement awaits computers or machines that can process and reason similar to the capabilities of human beings.

Android: *(an´droid) adj.* A robot or machine having a human appearance, anthropomorphic robot - i.e. a robot that looks like a human, such as some physical qualities resembling a person. *n.* An automaton that is created from biological materials and resembles a human. Many android developers call their creations "humanoids" rather than androids. Some androids are popularized by films. One example is Commander Data on *Star Trek: The Next Generation*.

Artificial Intelligence (AI) and Expert Systems:

The standard definition of artificial intelligence is "the ability of a robot or computer to imitate human actions or skills such as problem solving, decision making, learning, reasoning, and self-improvement" (Computers - The future of computers. Compton's Interactive Encyclopedia© 2000. [CD-ROM]. The Learning Company and its affiliates and licensors). Today's computers can duplicate some aspects of intelligence: for example, they can perform goal-directed tasks (such as finding the most efficient solution to a complex problem), and their performance can improve with experience (as with IBM's Deep Blue, chess-playing supercomputer). However, the programmer chooses the goal, establishes the method of operation, supplies the raw data, and sets the process in motion. Computers today are not, in themselves, intelligent.

Buckytubes: (buh´kee toobs, tyoobs) n. (also spelled as "buckeytube"; the spelling "buckytubes" is found more often) Single walled nanotubes (nanotubes literally means nano-sized tubes). They are a form of carbon based fullereness. One-billionth of a meter in diameter, these devices possess the electrical conductivity of copper, the thermal properties of diamond, and a tensile strength 100 times that of steel. First observed in the 1960's by Physicist Roger Bacon. Sumo Lijima, in 1992, using his NEC research team, perfected a way of making bulk quantities of single walled nanotubes. The name of buckytube is derived from architect Buckminster Fuller, and also applied to "buckyballs" - hollow and spherical, reminiscent of the geodesic domes by his design.

Closed-loop Control: A method of controlling robots in which the position of a moving member or the status of some other output is fed back to the control unit. A household thermostat is an example of a simple feedback loop. See also Feedback loop; Servocontrol.

DeoxyriboNucleic Acid (DNA): (dē ok´si ri´bō nōō—klē´ik) n.

A nucleic acid that carries the genetic information in the cell and is capable of self-replication and synthesis of RNA. DNA consists of two long chains of nucleotides twisted into a double helix and joined by hydrogen bonds between the complementary bases adenine and thymine or cytosine and guanine. The sequence of nucleotides determines individual hereditary characteristics.

End Effectors: A gripper or other tool attached to the wrist of a manipulator arm. This term usually applies to robotics.

Feedback Loop: A control system incorporating controller commands to a mechanism, the mechanism's response, measurement of the response and transmission of the measurement back to the controller for comparison to the original command.

Fitt's Law: (properly pronounced "fitzez law"),In 1954 Professor Paul Fitts of Ohio State University developed a model of human movement based on rapid, aimed movement. The law reads as, "The time to acquire a target is a function of the distance to and size of the target." It is an effective method of modeling rapid, aimed movements, where one appendage (like a hand) starts at rest at a specific start position, and moves to rest within a target area. Mathematically, Fitt's Law is stated as follows: $MT=a+b\ log2(2\ A\ /\ W)$, MT = movement time, a, b = regression coefficients, A = distance of movement from start to target center, W = width of the target.

Fullerene: (fo͞ol′ə rēn′) n. Fullerenes are large carbon-cage molecules. Natural carbon can exist in several forms. Graphite and diamond are the more widely known, but there is a third type - fullerenes. Fullerenes have been found to exist in interstellar dust as well as in geological formations on Earth. Fullerene cages are about 7-15 angstroms in diameter (that's about a billionth of a meter, or 6-10 times the diameter of a typical atom). In atomic terms, they are enormous.

High-Capacity Sensors: A robot sensor consists of a transducer, which converts physical properties into a signal (e.g. electrical) and a processor, which transforms the signal into the information required to perform a given task.

Industrial Robot: A reprogrammable manipulator designed to perform functions in industry normally performed by a person, and to perform those functions in essentially the same manner and at approximately the same speed as a person (Susnjara, Ken. A Managers Guide to Industrial Robots. Corinthian Press. 1982).

The Robot Institute of America (RIA) defines an industrial robot as: "A reprogrammable, multi-functional manipulator designed to move material, parts, tools, or specialized devices, through variable, programmable motions, for the performance of a variety of tasks." In non-industrial applications, the term may be applied to any mechanical device that is able to perform some task or movement under automatic control.

Lithography: (li thog´ rə fē) n. Manufacturing method using subtractive means in conjunction with chemistry or mechanics to etch away materials. Currently silicon is etched away to create gears and other machinery to operate micromotors.

Machine Vision: A device that can collect data and form an image that can be interpreted by a robot computer to determine an approximate position or to "see" an object.

Macroscopic: (mak′rə skop′ik) *adj*. Can be seen with the unaided eye. Relates to observations with the untrammeled eye. Opposite of Microscopic.

Meter: (mē′ tər) *n*. The base SI [Systeme International] unit of length. In 1983 the General Conference on Weights and Measures decided that the accepted value for the speed of light would be exactly 299,792,458 meters per second, so the meter is now defined as the distance traveled by light in a vacuum in 1/299,792,458 second. One meter is equal to about 39.37 in. in the U.S. Customary System of Measurements.

Metre: (mē´ tər) *n.* *Chiefly British.* variant of meter.

See **Meter** for definition.

Micromachining: This generally refers to machines of smaller sizes, usually microns, or millionths of a meter, in size; for example 1 millimeter equals 1,000 microns. Sometimes referred to as Micro-Electro-Mechanical-Systems (MEMS).

Microscopic: (mi´krə skop´ik) *adj.* Too small to be seen with the unaided eye but can be seen with a microscope. Relates to a microscope. Opposite of Macroscopic.

Moore's Law: (properly pronounced morz law) prov. The observation that the logic density of silicon integrated circuits has closely followed the curve (bits per square inch) = 2^(t - 1962) where t is time in years; that is, the amount of information storable on a given amount of silicon has roughly doubled every year since the technology was invented. This relation, first uttered in 1964 by semiconductor engineer Gordon Moore (who co-founded Intel four years later) held until the late 1970s, at which point the doubling period slowed to 18 months. The doubling period remained at that value through time of writing (late 1999). Moore's Law is apparently self-fulfilling. The Jargon Dictionary [World Wide Web] Retrieved June 20, 2004 from the World Wide Web: http://info.astrian.net/jargon/terms/n/nanobot.html.

Nano- : (nan´oh) n. Extremely small; (Greek: dwarf), one billionth (10^{-9}). A nanometer is one-billionth of a meter.

Examples as applied to Nanotechnology:

1)Ten hydrogen atoms lined up would fit within a nanometer.

2) Our DNA molecules are 2.5 nanometers wide.

3) A typical bacterium, say E. coli, is a thousand times bigger, measuring between 1,000 and

2,000 nanometers, while a virus like the ones that cause the common cold measures around 20 nanometers.

4) The width of the dot above this letter j is approximately 1 million nanometers.

5) Ten nanometers is one thousand times smaller than the diameter of a human hair.

Nanobot: *(nan′ oh bot) n.* A robot of microscopic proportions, presumably built by means of nanotechnology. As yet, only used informally (and speculatively!). Also called a `nanoagent'.

The Jargon Dictionary

[World Wide Web] Retrieved June 20, 2004 from the World Wide Web: http://info.astrian.net/jargon/terms/n/nanobot.html.

Nano-scale Robotic Arm: Precise positioning of molecules using molecular-scale devices and one-ten-millionth-scale robotic arms that could systematically add or withdraw atoms by sweeping back and forth over a surface.

Nanotechnology: (nan´ə tek nol´ə jē) n. The art of manipulating materials on an atomic or molecular scale especially to build microscopic devices, such as robots. A nanometer is one-billionth of a meter.

NASA: National Aeronautics and Space Administration. An independent agency of the U.S. Government responsible for the Civil Space Programme. It was established in 1958 by President Eisenhower based on the old National Advisory Committee for Aeronautics (NACA). It's head-quartered in Washington, D.C., where programme plans originate. Individual projects are implemented in different fields centers. Kennedy Space Center in Cape Canaveral, Florida is the center for launch operations.

Parallel Processing Computers: Incomputers, parallel processing is the processing of program instructions by dividing them among multiple processors with the objective of running a program in less time. In the earliest computers, only one program ran at a time. A computation-intensive program that took one hour to run and a tape copying program that took one hour to run would take a total of two hours to run. An early form of parallel processing allowed the interleaved execution of both programs together. The computer would start an I/O operation, and while it was waiting for the operation to complete, it would execute the processor-intensive program. The total execution time for the two jobs would be a little over one hour.

Real-Time Asset Tracking: Refers to the actual time during which the physical processing or operation takes place; also, the time logged on a robot in which it is or could be performing work.

Robot: The Robot Institute of America (RIA) defines *robot* as: "A reprogrammable, multi-functional manipulator designed to move material, parts, tools, or specialized devices, through variable, programmable motions, for the performance of a variety of tasks." In non-industrial applications the term may be applied to any mechanical device that is able to perform some task or movement under automatic control (Hall, Ernest L., Bettie C. Hall. Robotics: A User-Friendly Introduction. CBS College Publishing, 1985).

Robotics: The study of robots; the science of designing , building, and applying robots. Isaac Asimov is credited as the first to use the word *robotics* in the March 1942 issue of *Astounding Science Fiction*.

Scanning Tunneling Microscope (STM):

An instrument capable of creating atomic resolution images of the surface of electrically conducting materials.

Service Robot: According to The International Federation of Robotics (IFR) in 1997: "A service robot is a robot that operates partially or fully autonomously to perform services useful to the well being...of humans and equipment. They are mobile or manipulative or a combination of both."

A further definition from the Fraunhofer Institute for Manufacturing Engineering and Automation (IPA) in 1994: "A service robot is a freely programmable mobile device carrying out services either partially or fully automatically. Services are tasks that do not serve the direct industrial manufacture of goods, but the performing of services for humans and equipment."

Speech Recognition: The ability of a robot to communicate or perform tasks by the use of a spoken language. Not only understanding the words but also their meanings and implications.

Transducer: Device to convert energy from one form to another (examples include an electronic pressure gauge which converts mechanical pressure into electrical signals and a photocell which converts light into electrical signals).

BIBLIOGRAPHY

Asimov, Isaac. Karen Frenkel. Robots: Machines in Man's Image. Harmony Books. 1985.

Caudill, Maureen. In Our Own Image: Building an Artificial Person. Oxford University Press. 1992.

Cone, Robert J. How The New Technology Works [a guide to high-tech concepts]. Oryx Press. 1998.

Dorf, Richard C. Robotics and Automated Manufacturing. Reston Publishing Company, Inc. 1983.

Fouke, Janie, Editor, Trudy E. Bell and Dave Dooling, Writers. Engineering Tomorrow: Today's Technology Experts Envision the Next Century. IEEE Press Marketing. 2000.

Fjermedal, Grant. The Tomorrow Makers: a brave new world of living-brain machines. Macmillan Publishing Company. 1986.

Graham, Ph.D., P.E., James H. Editor. Safety, Reliability, and Human Factors in Robotic Systems. Van Norstrand Reinhold, 1991.

Hall, Ernest L., Bettie C. Hall. Robotics: A User-Friendly Introduction. CBS College Publishing. 1985.

Hunt, V. Daniel. Smart Robots: A Handbook of Intelligent Robotic Systems. Chapman and Hall. 1985.

Morik, Katharina, Michael Kasier, and Volker Klingspor (Eds.). Making Robots Smarter: Combining Sensing and Action Through Robot Learning. Kluwer Academic Publishers. 1999.

Nell W. Meriwether. Successful Research Papers in 12 Easy Steps. National Textbook Company Publishing Group. 2000.

Regis, Ed. Nano the emerging science of technology: remaking the world - molecule by molecule. Little, Brown and Company. 1995.

Schraft, Rolf Dieter and Gernot Schmierer. Service Robots: Products, Scenarios, Visions. A K Peters, Ltd.. 2000.

Susnjara, Ken. A Managers Guide to Industrial Robots. Corinthian Press. 1982.

Timp, Gregory (Ed.). Nanotechnology. AIP Press - Imprint of Springer-Verlag New York, Inc. 1998.

WEB SITES

A Debate About Assemblers [World Wide Web] Retrieved March 22, 2004 from the World Wide Web: http://www.imm.org/ sciamdebate2/ smalley.html

Feynman Grand Prize [World Wide Web] Retrieved March 3, 2004 from the World Wide Web: http://www.foresight.org/grand prize.1.html

Foresight Debate with Scientific American; Foresight Institute vs. Scientific American; Debate on Nanotechnology: Round 1 from Merkle. [World Wide Web]. Retrieved March 22, 2004 from the World Wide Web: http://www.foresight.org/ sciamdebate/sciam response.html

Kismet, the robot. [World Wide Web] Retrieved March 27, 2004 from the World Wide Web: http:// www.ai.mit.edu/projects/ sociable/baby-bits. html

Robots - Hans Moravec [World Wide Web] Retrieved March 25, 2004 from the World Wide Web: http://www.pbs.org/wgbh/nova/ robots/ moravec.html

Zyvex - Nanotechnology [World Wide Web] Retrieved March 3, 2004 from the World Wide Web: http://www.zyvex.com/nano

2001 Pocket Guide to Transportation: Chapter 3: Mobility: Table 7, Number of Aircraft, Vehicles and Vessels Mode: Passenger cars 1998 [World Wide Web] Retrieved March 25, 2004 from the World Wide Web: http://www.bts.gov/ publications/pocketguide

About the Author

Jon D. Sutherland has a Bachelor of Science degree in Mechanical Engineering from Kennedy-Western University, in Cheyenne, Wyoming. Most of his work experience has been in the field of Automotive and Plastic Extrusion specializing in machine and gauge design. His interests have always led him to look for new and innovative engineering technologies. He lives in Abbeville, South Carolina.

Printed in the United Kingdom
by Lightning Source UK Ltd.
103730UKS00001BA/7-15